World War II
Nose Art
In Color

Jeffrey L. Ethell

Motorbooks International
Publishers & Wholesalers

First published in 1993 by Motorbooks International Publishers & Wholesalers, PO Box 2, 729 Prospect Avenue, Osceola, WI 54020 USA

© Jeffrey L. Ethell , 1993

All rights reserved. With the exception of quoting brief passages for the purpose of review no part of this publication may be reproduced without prior written permission from the Publisher

Motorbooks International is a certified trademark, registered with the United States Patent Office

The information in this book is true and complete to the best of our knowledge. All recommendations are made without any guarantee on the part of the author or Publisher, who also disclaim any liability incurred in connection with the use of this data or specific details

We recognize that some words, model names and designations, for example, mentioned herein are the property of the trademark holder. We use them for identification purposes only. This is not an official publication

Motorbooks International books are also available at discounts in bulk quantity for industrial or sales-promotional use. For details write to Special Sales Manager at the Publisher's address

Library of Congress Cataloging-in-Publication Data

Ethell, Jeffrey L.
    World War II nose art in color / Jeffrey L. Ethell.
       p. cm. — (Enthusiast color series)
    Includes Index.

ISBN 0-87938-819-6
1. Airplanes, Military—United States—Decoration—History. 2. World War, 1939-1945—Aerial operations, American.   I. Title. II. Series.
UG1240.E86 1993
358.4'183'097309044—dc20   93-13072

Printed and bound in Hong Kong

On the front cover: Maj. Byron Trent stands in front of *Just Once More* one of the B-17s under his command when he was commander of the 333rd Squadron, 94th Bomb Group at Bury St. Edmonds, England, in 1945. The Vargas pose had been much copied in many different degrees of undress since it first appeared as a September 1943 *Esquire* calendar page. *Byron Trent*

On the frontispiece: *Hookem Cow* gets some care on its 755th Squadron revetment at Horsham St. Faith, England, in August 1944. The 458th Bomb Group flew 240 missions from February 1944 through the end of the war, but *Hookem Cow* didn't make it. She crashed on takeoff on 14 April 1945 as the Eighth Air Force headed out to bomb the remnants of Nazi Germany's war making potential. *NASM*

On the title page: After its combat tour in the 491st Bomb Group, *Tubarao* was painted with yellow and dark green stripes on its wings and fuselage as the group's assembly aircraft. The idea was to make the B-24 gaudy enough not to miss as the bombers took off from North Pickenham and joined up before heading across to targets on the Continent. *Robert Astrella*

On the back cover: Crewman Albert Krassman stands next to the nose art on the B-24 *Kentucky Belle. Albert R. Krassman*

Contents

Introduction 7
Nose Art Gallery 9
Index 96

Introduction

During Operation Desert Storm the unique phenomenon of decorating aircraft with personalized artwork reached a zenith. Though nose art had been reborn during the 1980s, peacetime does not provide the ideal climate for survival of the genre. It usually takes a war. By the time the short conflict with Iraq ended in early 1991, almost every flying machine in the theater was decorated with fanciful icons—some cute, some raunchy. When the aircraft returned home, the nose art disappeared almost overnight with the usual official disapproval, much to the disgust of the crews.

Much inspiration for Desert Storm nose art came directly from World War II, the heyday of this unique expression of military life. Pilots and crews wanted to emulate the previous generation who had used metal skin for canvas all over the world. Though some color photography from World War II has been available, on the whole readers have had to be content with looking at a black-and-white photo, then trying to fill in the color mentally.

Slowly but steadily, that is becoming less the case as new collections of wartime Kodachrome surface from GIs who were there with the foresight to use Kodak's new slow-speed transparency film. As a result, this is the first book on World War II nose art consisting solely of vintage color. In my previous Motorbooks Interna-

Opposite page
Punkie II, 5Q-O, was flown by 1st Lt. Vern Blizzard in the 504th Squadron, 339th Fighter Group, out of Fowlmere, England. The name came from Blizzard's nickname for his wife. From April 1944 to April 1945, the 339th racked up claims for 239.5 air and 440.5 ground victories, the highest of any unit for a single year in combat, including over 105 on 5 April 1945 and 118 on 11 April 1945. It was the only group to get over 100 strafing kills in two different missions. *R. Vern Blizzard*

R. T. Smith stands in front of his first *Barbie*, named for his wife, after returning to the US after a tour with the American Volunteer Group, the Flying Tigers. After flying the P-47 and getting reacquainted with the Army Air Forces, R. T. went back to China and Burma to fly P-51s and B-25s with the 1st Air Commando Group in 1944. He then came back home to serve as ops officer in a P-38 gunnery school at Van Nuys, California. *R. T. Smith*

tional volumes, *Fighter Command* and *The History of Aircraft Nose Art* (both published in 1991), I told the nose art story in great detail, from origins to interviews with the artists themselves, much of it illustrated with wartime Kodachrome, so there is no need to repeat it.

The goal of this book is to add significantly to the visual record using fresh material—but the primary idea is to have fun without taking the subject too seriously. Unfortunately the majority of people who painted nose art in World War II remain obscure, so building on what I have written in the other two books becomes harder each year. They should be interviewed, but the odds of finding them are low. All we can do is enjoy what they created and pass on the legacy to future generations.

My deepest thanks to those listed in the credits for each photo. Due to the efforts of these few, our historical memory is slowly changing from leaden shades to brilliant bloom.

Nose Art Gallery

During World War II, nose art often followed the GI maxim, "if it moves, salute it, if it doesn't, paint it." This applied to everything from motorcycles and jeeps to the largest bombers, as noses and other parts of aircraft were painted with personal decoration. Though the AT-6 was left behind when the pilots went off to war, this advanced trainer showed up with numerous squadrons throughout the world for instrument-currency training and general "hack" flying, even joy riding. *Bridget* was attached to the 7th Photo Group, which carried blue spinners on its Mustangs and Lightnings, at Mt. Farm, England, where she is getting some attention. Many enlisted men got their only chance to fly in hack AT-6s. *Robert Astrella*

Crew Chief Wally Carrier stands next to his 334th Squadron Mustang, *Zoomin' Zombie*, QP-I, which he, on occasion, loaned to pilot William J. Dvorak. The nose art carries the distinctive style of the group's premier nose artist, Don Allen, whose creations became a part of the 4th's identity. He was always in great demand, spreading his talent across a wide variety of aircraft. *Donald W. Allen*

Left
The 78th Fighter Group was one of the Eighth Air Force's premier outfits, having flown its first mission with Thunderbolts in April 1943. After 450 missions it had accumulated 338.5 air and 358.5 ground kills. It made a record 135 strafing kills on 16 April 1945 and claimed the first Me 262 jet fighter on 28 August 1944. *Mr. Ted III* was the mount of group commander Col. Frederic Gray. *Robert Astrella*

Covered with Synthetic Haze paint, *School Boy Rowe* was flown by 22nd Squadron, 7th Photo Group pilot Lt. David K. Rowe out of Mt. Farm, England. The light blue paint, much like RAF Photo Recon Unit (PRU) Blue, had the uncanny effect of making the aircraft seem to disappear at altitude. This didn't help Rowe at all when he was flying out of Poltava, Russia. Apparently breaking some rule or mistaken as an Fw 189, he was jumped by a Russian P-39 Airacobra and shot down. Fortunately, he bailed out safely. *Robert Astrella*

Left
When 7th Photo Group pilot Lt. Claude Murray flew F-5C Lightning *Dot+Dash* from Mt. Farm, England, on 6 October 1944 he had no idea neither would be coming back. Normally assigned to Lt. Everett Thies, *Dot+Dash* was taken by Murray, who was high over Germany when an Me 262 jet fighter bounced him. As Murray recalled, the Lightning "had been hit and I didn't know where. A huge, huge impact, a big jolt shuddered through my body. Something had happened! My right engine burst into flames." Limping back toward the Channel, Murray passed Arnheim and was over the Zuider Zee when smoke flooded the cockpit. He rolled over, dropped out of the stricken F-5, and parachuted into Holland, where he evaded the Germans for seven months with the aid of the Dutch underground. *Robert Astrella*

The F-15 Reporter was the photo recon version of the XP-61E Black Widow night fighter. As evident on *Klondike Kodak*, the F-15 had an excellent camera bay in place of the fighter's armament. The two place machine had an incredible range of almost 4,000 miles, giving it the ability to roam vast expanses at will, though the type did not see action in World War II. *Ole C. Griffith*

Right
The extensive blue camouflage paint on this 7th Photo Group Lightning at Mt. Farm provided a near perfect backdrop for this reproduction of a Varga Girl and was a tremendous aid in avoiding enemy fighters, though the 7th had to bring in P-51 Mustangs to fly cover for them toward the end of the war. The "Photo Joe" was often a frustrated fighter pilot who dreamed of having guns instead of film to shoot. In reality, the photo-recce mission proved to be one of the most crucial of the war as commanders came to depend increasingly on visual intelligence to move the Allied armies forward. *Robert Astrella*

William Fowkes stands in front of his P-38 Lightning, *Billy's Filly*, which he flew with the 12th Squadron, 18th Fighter Group out of Zamboanga, Philippines in early 1945. At that point in the war, the Lightning was the Army Air Forces' major fighter in the Pacific, the mount of both Dick Bong and Tom McGuire, the leading American aces of all time with 40 and 38 kills respectively. Its long range, ability to come home on one engine, and excellent grouped firepower endeared it to pilots in the Pacific. *William Fowkes*

After the 318th Fighter Group received 30 well-used P-38s from the 21st Fighter Group for B-24 escort missions, it was not uncommon to have a number of mechanical problems. *All Bugs!* certainly tells some of the story. The "Lightning Provisional Group" did a good job over Iwo Jima, Truk, and other distant targets until newer P-47Ns arrived in early 1945 to carry on the job. *James Weir*

Lt. Ed Kregloh on the day of his 100th mission with the Twelfth Air Force's 346th Squadron, 350th Fighter Group at Pisa, Italy, in March 1945. Disney's cartoon character Goofy makes his way though flak in this fanciful rendition which appeared on many 346th Thunderbolts. *Ed Kregloh*

T/Sgt. Thomas Dickerson cleans the guns on *California Cutie*, a 55th Squadron, 20th Fighter Group P-38J, code KI-S, flown by 1st Lt. Richard O. Loehnert. The white band and polished nose were meant to confuse enemy pilots on whether the Lightning was a Droop Snoot or not. The mission symbols stood for many different things: locomotives for those claimed in strafing, bombs for fighter-bombing missions, brooms for fighter sweeps, umbrellas for top cover, top hats and canes for bomber escort. *John W. Phegley*

Left
Lt. Hal H. Dunning sits on the wing of *Smokepole*, his 19th Squadron, 318th Fighter Group P-47D, which carried him through some rugged times. When bombing and strafing Pagan Island in August 1944 he "failed to notice a line of coconut trees until pulling out of my dive. There was very little time for decision. I chose to pull up straight and hit the top of one of the trees. After rejoining, Bill Loflin looked me over and reported no apparent damage. The aircraft flew normally.... My crew chief, Sgt. Eldo Lau, reported that he removed about 16 bushels of leaves, twigs, and branches from the engine and the turbocharger duct." *Paul Thomas/Bob Rickard*

Red-E Ruth was a 19th Squadron, 318th Fighter Group P-47 flown by Lt. Leon Cox, who destroyed three Oscars in short order on 25 May 1945. The Japanese pilots never did a thing to maneuver, though some of the Val bombers that the Oscars were there to cover made passes at the P-47s and maneuvered violently in spite of carrying bombs. The Thunderbolts had no trouble in downing the entire flight. *Paul Thomas/Bob Rickard*

Below
An apt description of the nose art, the name of this 333rd Fighter Squadron P-47N was *Too Big, Too Heavy*, a double meaning which also referred to the Thunderbolt, the largest American single-engine fighter of the war. The squadron's flying cobra emblem was painted on most of the unit's P-47s, a practice common throughout the 318th Fighter Group's individual squadrons. *via Ed Wolak*

Short Snorter was the other side of the 333rd Squadron's *Too Big, Too Heavy*. The similar nose-art style is more than evident. *James Weir*

This 9th Squadron, 49th Fighter Group P-38, *Irish*, sits in the dust of an airstrip in New Guinea. The 49th was the first fighter group to enter combat in the South Pacific, at war from early 1942 in Australia to the end, seeing a tremendous amount of action, flying P-40s, P-38s, and P-47s. Every pilot in the group had his good and bad comments on each type, but only the 9th Squadron flew P-38s, remaining extremely loyal to the Lightning, and getting 276 of the group's 678 victories. *Thomas K. Lewis via Vivian Lewis*

Right
2nd Lt. John L. Lowden stands next to one of the Waco CG-4A gliders he flew in combat with the 98th Squadron, 440th Troop Carrier Group. The unit was towed out of Orleans, France, for Operation Varsity, the invasion of Germany, landing near Wesel on 24 March 1945. Not since 1806 and Napoleon's invasion had an enemy force breached the Rhine River. Casualties the first day were heavier than during the initial assault of occupied France on D-day. Flying combat gliders was a very dangerous job with 25 percent casualties taken as normal. *John L. Lowden*

I n many cases the painting had a special meaning to the pilot and it served to keep his spirits up—after all, they were in a mighty rough business.

—Don Allen, nose artist, 4th Fighter Group

The US Navy was more than strict about not allowing any form of personal art to be painted on their aircraft (though this broke down a bit in the Pacific during combat), but the Royal Navy had no such problems. This Corsair Mk.I is being readied for an acceptance test flight by a Fleet Air Arm (FAA) pilot at NAS Quonset Point in September 1943. The first FAA Corsair unit, No. 1830 Squadron, was commissioned on 1 June 1943, and attached to HMS *Illustrious*. *National Archives via Stan Piet*

Right
Capt. Fred J. Christensen sits in his 56th Fighter Group P-47D, marked as the victor over 22 German aircraft. Christensen was assigned to the group in August 1943, becoming the first Eighth Air Force pilot to down six enemy aircraft on a single mission, performing the amazing feat on 7 July 1944. The Thunderbolt was named *Miss Fire*, and *Rossi Geth II*, after his girlfriend Rosamond Gethro. Zemke's Wolfpack, nicknamed because of the 56th commander Hub Zemke's talented leadership, came right down to the end of the war in a race with the 4th Fighter Group for top-scoring honors. Though it had more fighter aces than any other USAAF group, and more air-to-air kills (664.5), the final score was 992.5 for the 56th and 1,019 for the 4th, the highest of any USAAF unit in the war. *NASM*

> I didn't know anything about solvents, so I sometimes used 100-octane fuel.
>
> —*Arthur De Costa, nose artist, 355th Fighter Group*

Bombardier Capt. James G. Ratterree stands in front of plane No. 371, a 498th Squadron, 345th Bomb Group B-25 that carried a modified Falcon squadron insignia as nose art. This Mitchell was usually flown by Capt. Earl L. Giffin. *John P. Bronson*

Skirty Bert III was a P-47D in the 514th Squadron, 406th Fighter Group during operations on the European continent in 1945. Forward-base conditions were often more than rough, with bitter cold, constant rain or snow, and ever present mud. The 406th, in the forefront of the fighting with the Ninth Air Force, was the first unit to use five-inch HVAR air-to-ground rockets, which could do a mean job on German tanks. *John Quincy via Stan Wyglendowski*

Sharkmouths have long been celebrated on aircraft, but none were more famous than those on the Curtiss P-40. When the company built its 15,000th fighter, a P-40N, they painted the famous shark's teeth on the snout, as had so many P-40 units since No.112 Squadron, RAF, and the Flying Tigers, but then took a massive leap in putting art on an airplane. In addition to the Curtiss-Wright logo just ahead of the windscreen, Curtiss decided to paint the insignia of every nation that had flown a Curtiss fighter, as well as the Disney-designed AVG tiger jumping through a V, and kill markings representing action against the Germans and the Japanese. *NASM*

Rat Poison Jr. was a new Douglas A-26 Invader assigned to the 386th Bomb Group in early 1945. Though the group was sad to give up its slick Marauders, pilots quickly fell in love with the hot new attack bomber, which had the cruise speed of a P-51 and excellent handling. The Invader would go on to fight in Korea and Vietnam. *Richard H. Denison*

Right
Sharkmouths on Mustangs were relatively rare, particularly in the European Theater of Operations, but they did show up when someone had the motivation, as seen on this 361st Fighter Group P-51 in England. *Al Simmons via Stan Piet*

Dirty Dora was flown by Vic Tatelman in the 499th Squadron, the Bats Out of Hell of the 345th Bomb Group. This B-25C entered combat with the 38th Group before being passed on to the the Air Apaches in August 1943. She went through the tough runs over Rabaul, Wewak, and Kavieng and after a year with the 345th had over 180 missions. The plexiglass nose has been painted over since .50 caliber forward firing guns have taken the place of the bombardier for low-level strafing. There were few more effective low level strafers than the B-25 and the Air Apaches. *T.S. Shreve via David W. Menard*

Left
Many of the 397th Bomb Group's B-26 Marauders were known for their striking shark mouths, as *Linda* clearly displays in mid-1944 while based in England. The Ninth Air Force's Marauders did a sterling job of low- and medium-level bombing on the Continent to hit German lines of communication while Allied troops advanced. By the time the war was over, the B-26, initially believed to be a crew-killer, had the lowest overall combat losses of any American aircraft. Its crews became fiercely loyal, knowing they had a fast, rugged, and capable machine. *Robert Astrella*

Left
The 12th Bomb Group had a series of similar distinctive fierce faces painted on their 75mm-cannon-armed Mitchells. This B-25H, *Sunday Punch*, was a good example. There's no doubt the bomber has been through some extensive action as evidenced by the .50 caliber side gun muzzle discharge on the side of the fuselage. *North American via Ernie McDowell*

Navigator Hank Redmond sits in *Blazin' Betty*, another of the 12th Bomb Group's uniquely painted cannon-armed B-25Hs at Fenny, India, in 1944. The armor plate beneath the pilot's window provided some extra protection, but it also was a source of continual trouble as water ran down behind it, producing extensive corrosion. Masking tape was applied around the seams on many aircraft but the stuff quickly blew away or dried out in the heat. *Hank Redmond*

Though there was nose art on stateside wartime aircraft, it was relatively rare and certainly more subdued than what could be found overseas. *Rhett Butler* was a B-17E attached to a Flying Fortress transition unit when its younger brother, the B-17F, was going off to the combat zones in late 1942. These older Forts got flown so much the control cables would stretch, making the controls sloppy and mushy, a real challenge to a fledgling bomber pilot. *NASM*

Previous pages
The 498th Squadron, 345th Bomb Group, had the falcon unit insignia quickly transferred to their B-25s. The strafer modification meant the bombardier's position was eliminated, and the entire plexiglass nose was painted over and their fierce birds were brushed on. Here Garvice D. McCall and crew stand in front of *Near Miss* at Dobodura, New Guinea, at the end of 1943. *John P. Bronson*

Right
The USAAF Sixth Air Force operated in the southern hemisphere during World War II, from Mexico through Central America and the Panama Canal, down to Chile and Argentina. As a result, it got virtually no attention compared to the more famous air forces in the combat theaters. Cpl. Alvin R. Schwartz, 10th Bomb Squadron mechanic at Howard Field, Panama, stands in front of his B-25, which carries the unit insignia for nose art. This was a common practice across the globe, though not as popular as personalized creations. The 10th flew its Mitchells throughout the Caribbean, from Puerto Rico to Trinidad to the Canal Zone in antisubmarine patrols and replacement aircrew training. *NASM*

Patched-up, war-weary B-25s like this one attached to the 14th Fighter Group—a P-38 Lightning unit—at Triolo, Italy in 1944, were often pivotal in keeping parts and supplies flowing to far-flung fighter units. Though often used for rest-camp trips or jaunts to some time off in larger cities like Rome, this Mitchell would make emergency ration and supply trips when official channels would bog down. The P-38 was a complex airplane, requiring more maintenance and parts than single-engine fighters, so the B-25 was launched off in search of necessary items on many occasions. *James Stitt*

Miss Nashville belonged to the 7th Photo Group at Mt. Farm, England. Originally assigned to the Eighth Air Force to help form a bomb group, the B-25C was turned loose when planners decided the type would never fly combat with US units in Europe. After some communications work, it was given to the 7th, painted all black and then flown on night recce missions using photoflash bombs. After being stripped to bare aluminum, she was used as a hack until destroyed by flak in October 1944 on a courier mission. *Robert Astrella*

Left
The crew of *Lonesome Polecat, Jr.* climbs aboard their 308th Bomb Group B-24 Liberator in China, late 1944. The 308th provided much of the heavy bombing power for Claire Chennault's Fourteenth Air Force. *NASM*

Three of the great combat pilots in the China-Burma-India Theater—R.T. Smith, Johnny Alison, and Phil Cochran—in front of Smith's 1st Air Commando Group B-25H, *Barbie III*, named for his wife. When Smith flew combat in the P-51A and B-25H he brought his experience as a Flying Tiger with him. Alison and Cochran were tasked with forming the 1st Air Commando to support Brigadier Orde Wingate's Chindits, a guerilla unit operating behind enemy lines. Cochran was made famous by cartoonist Milton Caniff as Flip Corkin in the comic strip *Terry and the Pirates*. *R.T. Smith*

Right
A veteran Air Apache B-25D strafer, *Old Schenley* was a popular 345th Group headquarters airplane normally assigned to the 501st Bomb Squadron. Group commander Col. Clinton U. True had his name painted under the pilot's window, but many ranking officers flew the bomber, often leading the group. She is seen here at Nadzab, New Guinea, with Dick Reinbold in the cockpit. *Dick Reinbold via Vic Tatelman*

M ost pilots had a basic idea of what they wanted and a name to go with it in most cases. I would make a small sketch, usually in black pencil but occasionally in color. Once the sketch was approved I'd scale it up directly on the plane. Somewhere along the way I established eight pounds Sterling (about $35) as the price for a paint job on a plane; five pounds if just lettering.

—*Don Allen, nose artist, 4th Fighter Group*

Ninth Air Force B-26 Marauders were famous for a large variety of nose art, some outrageous, some not. *Sure Go For No Dough* combined a takeoff on popular cartoon art with the ever present topic of sex. *Robert Astrella*

Special Delivery F.O.B. was flown by 1st Lt. Thomas K. Lewis, originally the B-25D's copilot, with the 501st Squadron, 345th Bomb Group, the Air Apaches. Here she sits at Nadzab, New Guinea in early 1944. As was typical of so many B-25s with retrofitted sidepack .50 caliber machine guns, an aluminum doubler had to be riveted onto the fuselage in front of the blast tubes as a try at keeping the skin from buckling after prolonged combat. The blast damage is certainly evident as it has removed much of the black paint on the doubler.
Thomas K. Lewis via Vivian Lewis

Left
A 441st Squadron, 320th Bomb Group Marauder and crew wait with their aircraft in the early morning before a mission from Sardinia in 1944. Many AAF units had a continual series of rough living conditions, from dust to mud and back, with a pyramid tent the most luxurious of hostels. *Joseph S. Kingsbury*

Pilot W. T. Boblitt with his 441st Squadron, 320th Bomb Group B-26 Marauder, *Little Sherry*. The bomb symbols are clear testimony to the bomber's combat record in the Mediterranean Theater of Operations, typical of the 320th's aggressive operations. The group was known as one of the most successful of Marauder units, making its way through a tough campaign from North Africa to Italy via Corsica and Sardinia. *Joseph S. Kingsbury*

The Mudhen was a veteran Eighth Air Force B-17F that survived a number of missions until made a hack airplane. All of the mission symbols have been painted out, along with several other former personal markings. *Robert Astrella*

Below
Scorchy II flew with the 359th Squadron, 303rd Bomb Group out of England in mid-1944. The nose art, as with so many wartime aircraft, was based on one of Alberto Vargas' *Esquire* calendar pages, in this case October 1944. This lady appeared on airplanes all over the world in 1944 and 1945. *Robert Astrella*

The 490th Bomb Group was known for some very stunning nose art, well represented here by *Looky Looky* of the 851st Squadron. *Arnold N. Delmonico*

Right
One war-weary Fort, *Alabama Exterminator II*, went through quite a bit of use before being relegated to a worn-out hack, as she appears here in England, mid-1944. Though the F-style clear bubble nose can deceive, it was added to this B-17E, AAF serial number 41-9022, after its combat days. The bomber was attached to both the 341st Squadron, 97th Bomb Group and the 546th Squadron, 384th Bomb Group before being turned out to pasture. The number of things stuck and patched onto her reflect the note on the ADF "football" antenna fairing: "Holmes Bastard Bomber." *Robert Astrella*

Queenie was a 91st Bomb Group combat veteran B-17G flying out of Bassingbourn, England—and there were at least 13 bombers in England with the same name, and another named *Queeny* to boot. Sorting through which belonged to which units was often a difficult task. *Robert Astrella*

The Fifth Air Force insignia decorates the towel covering part of *Atomic Blonde*, a B-24M with the 531st Squadron, 380th Bomb Group during the last part of the war as she sits on Okinawa in 1945. The Liberator, much loved by Fifth commander George Kenney, remained an essential part of the Army Air Forces' striking power against island strongholds while Twentieth Air Force B-29s took the war to the Japanese homeland. *John R. Trease via Larry Hickey*

Love 'em All was another 490th Bomb Group Fort. She flew 40 combat missions with the 849th Squadron by the time the war ended. *Arnold N. Delmonico*

Left
Lazy Lou saw her best days with the 446th Bomb Group flying out of Bungay, England. Here she's already managed to make it though enough missions to beat the odds. *Albert R. Krassman*

Assam, India, served as an almost inexhaustible source for a play on words painted on numerous aircraft, such as this CBI B-24 *My Assam Wagon*, one of the better examples. The most common version was some form of Assam and dragon. From the looks of this combat-worn veteran, no doubt the crew had some affection for the name. *via Campbell Archives*

Right
Al Keeler, second from right, and his 412th Squadron, 95th Bomb Group crew stand in front of their B-17G, which carried two different names: *GI Issue: Government Property* on the left side and *Sandy's Refueling Boys* on the right with a roll of toilet paper. The message was clear—they were simply expendable GI issue people, just like a roll of toilet paper. *Albert J. Keeler*

DRAGON LADY

Left
Milt Caniff's comic strips were more popular with servicemen than almost any other, and his creations appeared on aircraft throughout World War II. One of his most popular is reflected in *Dragon Lady*, though a bit more clothed than usual, when she flew with the 704th Squadron, 446th Bomb Group out of Bungay, England. *Albert R. Krassman*

Based with the 446th Bomb Group at Bungay, England, *White Lit'nin* was a 705th Squadron B-24H that ended its days in a salvage dump after a belly landing in March 1945. *Albert R. Krassman*

Of all the 490th Bomb Group's B-17s, *Carolina Moon* flew more missions than any other Fort, a total of 78. In addition to the mission symbols, the names "Adeline" and "Suzy" appear over the bombardier and navigator windows. It was not unusual for crew members to put the names of their favorites over their crew stations. *Arnold N. Delmonico*

Right
Nose art on late-model B-24s, such as this 446th Bomb Group Lib, could be painted smack dab in front. What it meant was another question. *Albert R. Krassman*

> During my military career I found that drawing cartoons was a great way to get out of some duty.
> —*David R. Hettema, nose artist, 91st Bomb Group*

Cartoon characters of the era were extremely popular as nose art subjects, particularly Walt Disney and Warner Brothers creations. Bugs Bunny was one of the more copied of Warner's stable. His wise-guy attitude and famous Mel Blanc voice suited GIs just fine. This 490th Bomb Group B-17G has captured Bugs to near perfection. *Arnold N. Delmonico*

Naughty Nan was another 446th Bomb Group Liberator, a B-24H with the 705th Squadron. She made it through quite her share of combat, as the bomb symbols attest, but was written off after a local hop and a forced landing at Bungay when the right landing gear leg wouldn't come down. *Albert R. Krassman*

Right
Though there may be several meanings to the nose art on this 705th Squadron, 446th Bomb Group Liberator, it clearly gives an indication that many crews felt they were really big children in a very dangerous place. Note the small flying helmet on top of the kid. *Albert R. Krassman*

It's hard to tell if *Peg o' San Antone* was a real person remembered by someone in the 706th Squadron, 446th Bomb Group, but there's no doubt she represented a great deal to many at Bungay, England. *Albert R. Krassman*

Right
Thar She Blows II sits on the dump at Saipan, abandoned after a significant combat record in the Seventh Air Force's 30th Bomb Group. There was a *III* to carry on the tradition and survive to return stateside. *Paul Thomas/Bob Rickard*

Dinky Duck was a popular comic character in the 1940s, as this tribute on a 706th Squadron, 446th Bomb Group B-24 shows. *Albert R. Krassman*

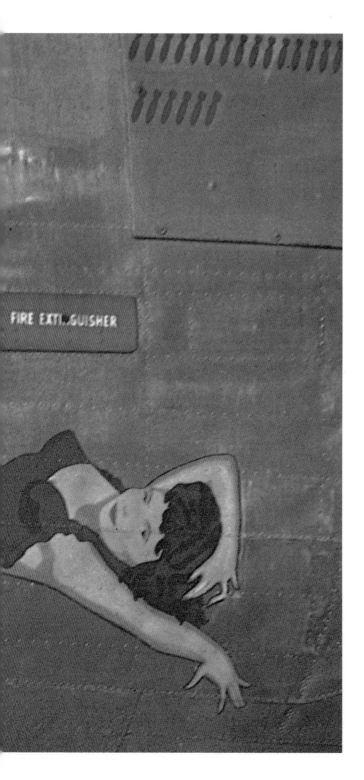

Another 446th Bomb Group Liberator, *Black M*. *Albert R. Krassman*

Left
Popular songs were strong subjects for nose art, but this one song seemed to dominate most of the others with at least twenty-seven examples in the Eighth Air Force alone. There were five examples of *Shoo Shoo Baby* in the 446th Bomb Group at Bungay, England, this one with the very popular Alberto Vargas *Esquire* gatefold. *Albert R. Krassman*

Kentucky Belle flew initially with the 490th Bomb Group before being transferred to the 446th Group at Bungay. The mission tally on the side includes 80 bombing missions, three trucks for carrying fuel to France, and a single supply drop. *Albert R. Krassman*

Left
The right side of the same airplane, *Kentucky Belle*, was quite a stunning example of nose art in the Vargas style, as crewman Albert Krassman's smile attests. *Albert R. Krassman*

This B-24H was one of the original Liberators with the 704th Squadron when the 446th Bomb Group went into action over Europe. In addition to the 49 mission symbols there is a swastika for claimed destruction of a German aircraft. Unfortunately she was so badly shot up by flak in October 1944 she had to be abandoned in France after a forced landing. *Albert R. Krassman*

Donald Duck went to war on all fronts, on all types of machines, from jeeps to bombers, as this 446th Bomb Group B-24 clearly shows. The addition of armor plate for the copilot has covered some of Donald's shoulder and head.
Albert R. Krassman

Right
Wistful Vista was a faithful reproduction of an *Esquire* magazine painting transferred to a 733rd Squadron, 446th Bomb Group Liberator.
Albert R. Krassman

Opposite page
Another 446th Bomb Group B-24, *Old Faithful*, represents American hitting power and patriotism, among the common themes for nose art. *Albert R. Krassman*

Left
Sugar Baby of the 446th Bomb Group warms up on her 704th Squadron hardstand at Bungay, England, in early 1945. *Albert R. Krassman*

Below
Satan's Little Sister rides into combat on the side of this 446th Bomb Group, 706th Squadron Liberator. *Albert R. Krassman*

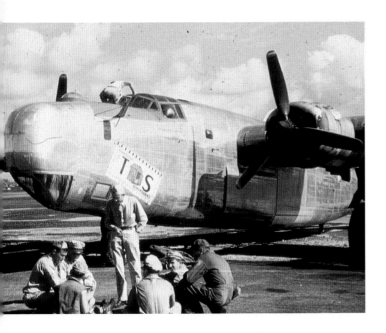

A 4th Photo Recon & Mapping Squadron Liberator on the ramp at Sydney, Australia has a play on words for condolences when things were tough. The stripped down bombers could make excellent speed at high altitude, the higher the better to avoid interception. *Pete Malone via Larry Hickey*

Ole Patches, an F-7 photo-recon version of the B-24, flew with the 4th Photo Recon & Mapping Squadron, Fifth Air Force, out of Hollandia, New Guinea, in April 1945. The very valuable, but long, mapping and recce missions flown by F-7s took them alone deep into enemy territory, and many were never heard from again. *Pete Malone via Larry Hickey*

Next page
Pom Pom Express, a B-24M of the 531st Squadron, 380th Bomb Group, 5th Air Force, sits on Okinawa in August 1945. *John R. Trease via Larry Hickey*

P ay? I got a lot of beer.
—*Al G. Merkling, nose artist, 20th Combat Mapping Squadron*

This C-47 in England was well named according to many crews who had to keep supply lines running, drop paratroops, or tow gliders. *Fatigue* exhibits some signs of wear, with chipping paint and panels off for maintenance. *Robert Astrella*

I started [*Taurus the Bull*] three times; then it would take off and never return.
—*Philip S. Brinkman, nose artist, 486th Bomb Group*

Right
When the 313th Bomb Wing settled into operations off Tinian, the local Seabees turned up with some very talented artists who painted at least sixteen pieces of nose art on wing Superforts, each representing a different Seabee unit and insignia. *National Archives via Stan Piet*

Previous pages
When the B-29 Superfortress began flying very long range missions to Japan proper, crews were stretched to their limits, suffering from exhaustion. The nose art on *Moonshine Raiders* gives a good representation of how they felt when they got back. *Wallace via Ernie McDowell/Larry Hickey*

Some Gooney Birds, though less in the news than their gun- and bomb-laden brothers, had elaborate nose art, like that of *Texas Hellcat* with the 433rd Troop Carrier Group. All four of the Fifth Air Force's C-47 troop carrier groups were placed under the 54th Wing. *Fred C. Baker via Larry Hickey*

Left
Numerous captured German aircraft were painted in bright, American hot-rod fashion, including this Fw 190 at the 386th Bomb Group's base at St. Trond, Belgium, spring 1945. When the bright-red Focke-Wulf was airborne in the local area there was no doubt about it being a friendly aircraft. *Richard H. Denison*

Left
Stud Duck flew with the Twelfth Air Force in a variety of resupply and troop-carrier missions. Donald Duck went to war in all forms and missions, here clearly as a man about town. *James Stitt*

The 313th Bomb Wing at North Field, Tinian, was fortunate enough to have the talented services of a Marine named Scott who painted numerous B-29s, including *Dina Might* of the 504th Group. *R.W. Teed via Fred Johnsen*

I did a rat in a Zoot Suit for Claiborne "Zoot" Kinnard, but I don't think we used it because it wasn't appropriate to Colonel Kinnard's command position.
—*Arthur De Costa, nose artist, 355th Fighter Group*

Once B-29s were operationally established overseas, some of the finest nose art of the war began to appear on them. The Superfortress' nose was an ideal canvas and talented artists made maximum use of it, evident from *Night Prowler*. *Wallace via Ernie McDowell/Larry Hickey*

Right
Bugger, a fanciful bombing dragon, goes off to war with the 331st Bomb Group, one of the last bomb groups to enter the war, flying its first combat mission on 9 July 1945 from Northwest Field, Guam. *Wallace via Ernie McDowell/Larry Hickey*

The Twentieth Air Force insignia was painted on several 500th Bomb Group B-29s, at Isley Field, Saipan. *73rd Bomb Wing Assn. via David W. Menard*

Left
The man most responsible for combat use of the B-29 was Maj. Gen. Curtis E. LeMay: his tactics put fear into the enemy—and into the Superfort crews—by sending the B-29s in low over the heart of Japan at night to ensure maximum fire-bomb effectiveness. Here, LeMay stands in front of a 6th Bomb Group B-29 on Guam on 14 April 1945. The unit's insignia was a rendition of pirate Jean Lafitte, which carried on a tradition from the group's days in the Panama Canal Zone during the 1920s and 1930s. No other artwork was allowed on the aircraft other than a red name, usually on a white streamer. *NASM*

Right
Holley-Hawk flew with the 58th Bomb Wing out of North Field, Tinian, in 1945 with an insignia "sponsored" by the 112th Seabees. *R.W. Teed via Fred Johnsen*

Tail Wind flew with the 73rd Bomb Wing out of Isley Field, Saipan. The wing was commanded by Brig. Gen. Emmett "Rosie" O'Donnell, one of the architects of Superfortress attack against Japan. When the war ended the B-29s had laid waste to 175 square miles of urban area in sixty-nine cities, leaving over nine million homeless, mostly due to the horrendous fire bombing raids carried out at low level. *73rd Bomb Wing Assn. via David W. Menard*

Left
A Seabee puts the finishing touches on 504th Bomb Group B-29 *Indian Maid*, one of the many 313th Wing bombers on Tinian that received the wonderful artistic talents of the Seabees, who adopted many of the aircraft. *National Archives via Stan Piet*

This 314th Bomb Wing flagship carried the insignia of each of its units with a running talley of missions flown and aircraft claimed destroyed during their stay at Northwest Field, Guam, in 1945. *Paul Grieber via Stan Piet*

I made friends with the natives, and that's how I got some of my colors.
—Rusty Restuccia, nose artist, 494th Bomb Group

The 497th Bomb Group's John G. Albright painted numerous pieces of nose art on the unit's bombers, including *Peace on Earth*, seen parked in its revetment on Isley Field, Saipan. She was forced to ditch on 4 March 1945 on the way home from Tokyo. Though Lt. Norman Westervelt did a great job of setting her down, he was washed off the wing as he scrambled out and was never found. Another of the crew was also lost. The other nine men were rescued by a Dumbo OA-10A Catalina, the most loved airplane in the Pacific. *73rd Bomb Wing Assn. via David W. Menard*

Another 6th Bomb Group Superfort, *Ernie Pyle's Milkwagon*, with Jean Lafitte on the nose. *via Frank F. Smith*

The 73rd Bomb Wing's *Mrs. Tittymouse* at Isley Field, 1945. *73rd Bomb Wing Assn. via David W. Menard*

Right
The Baroness was attached to the 73rd Bomb Wing on Saipan in 1945. *73rd Bomb Wing Assn. via David W. Menard*

Left
Lt. Francis Horne was given credit for 5.5 kills while flying *Snoot's Sniper* with the 352nd Fighter Group out of Bodney, England. The Mustang was supposed to have been named "Snoot's Snipper" in recognition of crew chief Art "Snoots" Snyder's ability as a barber. He painted the airplane as close to a barber pole as he could manage with red, white, and blue stripes on the trim tabs and around as many places as he could. The airplane became a flying advertisement for the "Snyder treatment." *Robert Astrella*

Below
Revetments of the 500th Bomb Group at sundown, Isley Field, Saipan, 1945. *73rd Bomb Wing Assn. via David W. Menard*

Previous pages
Mary Co-ED II flew with the 74th Squadron, 434th Troop Carrier Group through some of the toughest campaigns in Europe, as her mission tally attests. The unit took part in the airborne assaults on Normandy, Holland, and Germany and flew into the middle of Bastogne to resupply American soldiers cut off by the Germans in the Battle of the Bulge. This was more than dangerous because the Gooney Bird was simply a converted civil DC-3 airliner. It didn't even have self-sealing fuel tanks, which resulted in many going down in flames under enemy fire at low level. *Robert Astrella*

This American Southwest-Indian motif was popular in all theaters of war, as B-17F *Thunderbird*, one of at least six in England, shows. *Robert Astrella*

My Darling Dorothy was one of the 7th Photo Group's Spitfire PR.XIs at Mt. Farm, England. The deepened cowling housed an enlarged oil tank for long-distance flying, and additional fuel tanks were installed in the wings and fuselage. As a result, this short-range defensive fighter became a very long distance recce platform, able to fly almost unmolested at altitudes around 40,000 feet. Unfortunately it was cold, cramped, and noisy for the pilot wedged into the tiny cockpit. *Robert Astrella*

Index

20th Century Sweetheart, 83
Alabama Exterminator II, 46
All Bugs!, 16
Atomic Blonde, 48
Barbie, 8
Barbie III, 38
Billy's Filly, 15
Black M, 63
Blazin' Betty, 31
Bridget, 9
Bugger, 80
California Cutie, 19
Carolina Moon, 54
Dina Might, 79
Dinky Duck, 60
Dirty Dora, 29
Dot+Dash, 13
Dragon Lady, 53
Ernie Pyle's Milkwagon, 87
Fatigue, 74
GI Issue
    Government Property, 50
Goin' My Way, 57
Holley-Hawk, 83
Hookem Cow, 4
Indian Maid, 85
Irish, 22
Kentucky Belle, 65
Klondike Kodak, 14
Lazy Lou, 49
Linda, 30
Little Sherry, 43
Lonesome Polecat, Jr., 37
Looky Looky, 46
Love 'em All, 49
Mary Co-ED II, 94
Miss Fire, 24
Miss Nashville, 37
Moonshine Raiders, 74
Mr. Ted III, 11
Mrs. Tittymouse, 88
My Assam Wagon, 50

My Darling Dorothy, 95
Naughty Nan, 58
Near Miss, 34
Night Prowler, 80
Old Faithful, 69
Old Schenley, 38
Ole Patches, 70
Peace on Earth, 87
Peg o' San Antone, 60
Pom Pom Express, 70
Punkie II, 7
Queenie, 47
Rat Poison Jr., 29
Red-E Ruth, 21
Rhett Butler, 34
Rossi Geth II, 24
Sandy's Refueling Boys, 50
Satan's Little Sister, 69
School Boy Rowe, 13
Scorchy II, 45
Shoo Shoo Baby, 63
Short Snorter, 22
Skirty Bert III, 26
Smokepole, 19
Snoot's Sniper, 91
Special Delivery F.O.B., 42
Stud Duck, 79
Sugar Baby, 69
Sunday Punch, 31
Sure Go For No Dough, 41
Tail Wind, 85
Texas Hellcat, 77
Thar She Blows II, 60
The Baroness, 88
The Big Drip, 58
The Mudhen, 45
Thunderbird, 94
Too Big, Too Heavy, 21, 22
Tubarao, 4
White Lit'nin', 53
Wistful Vista, 66
Zoomin' Zombie, 11